THE CAVENDISH SQUARE ANIMAL LIBRARY

The World of ZEBRAS

BY JENNIFER LOMBARDO

Cavendish Square

Published in 2026 by Cavendish Square Publishing, LLC
2544 Clinton Street Buffalo, NY 14224

Website: cavendishsq.com

Cataloging-in-Publication Data

Names: Lombardo, Jennifer.
Title: The world of zebras / Jennifer Lombardo.
Description: Buffalo, NY : Cavendish Square Publishing, 2026. | Series: The Cavendish Square animal library | Includes glossary and index.
Identifiers: ISBN 9781502675248 (pbk.) | ISBN 9781502675255 (library bound) | ISBN 9781502675262 (ebook)
Subjects: LCSH: Zebras–Juvenile literature.
Classification: LCC QL737.U62 L596 2026 | DDC 599.665'7–dc23

Editor: Jennifer Lombardo
Designer: Deanna Lepovich

The photographs in this book are used by permission and through the courtesy of: Cover (image) nwdph/Shutterstock.com; cover and various pages (grass vector) Product Label/Shutterstock.com; various pages (tape element) Tamara Luiza/Shutterstock.com; p. 5 (top) 10 Cows Photos/Shutterstock.com; p. 5 (bottom) KIMFOREVERS/Shutterstock.com; pp. 6, 20 (top) EcoPrint/Shutterstock.com; p. 8 JohnatAPW/Shutterstock.com; p. 9 Quagga Naturhistorisches Museum Basel 27102013 3/Wikimedia Commons; p. 10 mhenrion/Shutterstock.com; p. 12 Peter Mullineux/Shutterstock.com; p. 13 WihanSaayman/Shutterstock.com; p. 14 Nicola_K_photos/Shutterstock.com; p. 16 JMx Images/Shutterstock.com; p. 18 Peter Hermes Furian/Shutterstock.com; p. 20 (bottom) MM.Wildlifephotos/Shutterstock.com; p. 21 (top) GranTotufo/Shutterstock.com; p. 21 (bottom) Arend van der Walt/Shutterstock.com; p. 23 Wirestock Creators/Shutterstock.com; p. 24 Andrzej Kubik/Shutterstock.com; p. 26 AndreAnita/Shutterstock.com; p. 27 Gunter Nuyts/Shutterstock.com; p. 28 paula french/Shutterstock.com; p. 29 GUDKOV ANDREY/Shutterstock.com; p. 31 Vintage_Space/Alamy Stock Photo; p. 32 WOLF AVNI/Shutterstock.com; p. 35 bluehand/Shutterstock.com; p. 36 CK-TravelPhotos/Shutterstock.com; p. 37 mishamartin/Shutterstock.com; p. 38 Hyracotherium Eohippus hharder/Wikimedia Commons; p. 40 Kletr/Shutterstock.com.

CPSIA compliance information: Batch #CSCSQ26: For further information contact Cavendish Square Publishing LLC at 1-877-980-4450.

Printed in the United States of America

CONTENTS

Introduction 4

Chapter 1: 7
Zebra Features

Chapter 2: 17
Habits and Habitat

Chapter 3: 25
Life and Death

Chapter 4: 33
Save the Zebras

Think It Through 42

Glossary 43

Find Out More 44

Index 46

About the Author 48

INTRODUCTION

Zebras look a lot like horses, and for good reason: The two species are related. Zebras are members of the horse family, and they have much in common with their relatives. They both live in groups called herds and can run very fast, which they enjoy doing in open fields. However, there are several differences between horses and zebras.

The most obvious difference is their coat. Experts believe the black and white stripes on zebras could help them escape predators. The theory, or idea, is that as the zebras run, the stripes make it confusing for creatures such as lions to pick out individuals among the herd. This confusion gives the zebras time to escape. Another theory is that the stripes help keep them cool in the hot sun. A third theory is that the stripes confuse biting flies, so zebras avoid being bitten.

Zebras are also smaller and more **aggressive** than horses. Some horses are skittish, meaning they are easily frightened or made nervous. A skittish horse may back away from someone trying to put a saddle on it or run when it feels threatened. Horses do sometimes bite, but this happens mostly when they are cornered and trying to defend themselves. In contrast, like many other wild animals, a zebra that feels threatened is likely to attack rather than run. They will attack people, predators, and even sometimes each other.

Despite their aggression, zebras are a favorite among wildlife enthusiasts. Their grace and beauty make them a delight to watch, and they play an important role in their local **ecosystem**.

Zebras are close enough to horses and donkeys that they can interbreed, or make babies together. A male zebra and a female horse make a zorse; a female zebra and a male horse make a hebra; and a donkey and a zebra make a zedonk or zonkey.

ZONKEY

ZORSE

Zebras can reach speeds of 30 to 45 miles (56 to 64 km) per hour.

CHAPTER 1

ZEBRA FEATURES

The most common and best known zebra is a species called the plains zebra. This type of zebra, as its name suggests, is found on the plains, mainly in eastern and southern Africa. However, there are two other species of zebra as well: Grevy's zebra and the mountain zebra. All three species look very similar, but they have some differences that make it possible to tell them apart.

PLAINS ZEBRA

The plains zebra has the widest stripes of the three species, and they are spaced farthest apart. The stripes extend onto the belly, and faintly onto the legs. There are six or seven subspecies of plains zebra. These include:

- Burchell's zebra
- Grant's zebra
- Selous's zebra
- Chapman's zebra
- Crawshay's zebra
- maneless zebra

THAT'S WILD!

Every zebra has a **unique** stripe pattern, the way humans have unique fingerprints. This is how zebra foals, or babies, find their mothers.

The seventh subspecies, the quagga, has been extinct since 1883. "Quagga" is the word for "zebra" in the language of the Khoikhoi people, an Indigenous, or Native, group from southern Africa. It is part of the Latin name for plains zebras: *Equus quagga*.

THAT'S WILD!

A subspecies is a group of a particular species that has traits that set it apart from other groups.

The maneless zebra, as its name suggests, has little to no mane.

THE QUAGGA PROJECT

The quagga was hunted to extinction when European colonizers moved into Africa. In 1987, a group of researchers in South Africa started trying to bring the quagga back. Since the quagga is a subspecies of plains zebra, they did this by **selectively breeding** certain zebras in an effort to change their stripe pattern to that of the quagga.

Scientists have tried many times to bring back extinct species, such as woolly mammoths. However, without the extinct species' exact DNA, this is impossible. However, because the quagga is a subspecies of zebra, it is the same as other zebras, with the only difference being how its coat looks. This means there is hope for the Quagga Project to succeed.

The quagga had the most unusual stripe pattern of all the zebras, as this museum example shows.

HOW THE ZEBRA GOT ITS STRIPES

This myth is told by the San people of Namibia.

Many years ago, when the world was new and so were the creatures in it, Africa had very little water. A baboon guarded one of the small pools and chased away anyone who got near "his" water.

One day a zebra and his son came to the pool to drink. The baboon yelled, "Get away! This water is for me only."

The zebra's son yelled back, "No! This water is for everyone."

The baboon responded, "There isn't enough for everyone. I'll fight to defend my water."

The two launched themselves at each other, and soon they were rolling around on the ground, biting and kicking. Finally the zebra's son was able to kick the baboon all the way up onto the rock cliffs around the pool of water. Tired and sore, the zebra's son staggered to his feet without looking where he was going, and fell right into the baboon's fire. The fire scorched his white fur with black stripes, and this is how zebras have looked ever since.

Although they fought in this story, zebras and baboons can often be seen drinking from watering holes together in real life.

Each subspecies is **subtly** different. For example, Burchell's zebra has what are called "shadow stripes" or "ghost stripes," which are fainter black stripes between the darker, thicker stripes. Selous's zebra has these as well, but they are generally darker than the ghost stripes on Burchell's zebra. Grant's zebra has no ghost stripes at all.

> **THAT'S WILD!**
> Because their sizes overlap, there is some debate over whether the plains or mountain zebra is the smallest species. Grant's zebra is the smallest and most common of all subspecies, so many people say plains zebras are the smallest.

The height of a zebra, like any other horse, is measured from the ground to its shoulder. Plains zebras are generally between 3.61 and 4.76 feet (1.10 and 1.45 m) tall and weigh between 386 and 849 pounds (175 and 385 kg).

MOUNTAIN ZEBRA

Mountain zebras measure between 3.81 and 4.79 feet (1.16 and 1.46 m) tall and weigh between 450 and 948 pounds (204 and 430 kg). There are two subspecies of mountain zebra: the Hartmann's mountain zebra and the Cape mountain zebra. While they each have features that help distinguish them from plains zebras, such as smaller ears and stripe patterns that go all the way down to their hooves, the fastest way to tell a mountain zebra from a plains zebra is by looking for the dewlap, which is a flap of skin on the neck. A mountain zebra has one; a plains zebra does not.

> **THAT'S WILD!**
> Mountain zebras' hearts weigh more than plains zebras'. Extra powerful hearts help pump blood at high altitudes, where there is less oxygen in the air.

A Cape mountain zebra's dewlap is clearly visible in this picture.

BLACK AND WHITE, OR WHITE AND BLACK?

Many people believe zebras are white with black stripes because the black stripes **taper** off near the belly and legs, leaving the rest white. However, they are actually black with white stripes!

Researchers have shaved small patches of hair off zebras and found that the skin underneath is black. However, this does not mean the black patches on zebras are not covered in hair. A zebra's hair contains cells that produce a **pigment** called melanin, which colors some hairs black. If there is no melanin in the hair, it is white.

The muzzle, or nose, of a zebra shows its black skin.

Experts are unsure of exactly what a dewlap is for, but there are theories. One is that it helps the animal keep cool. Another is that it attracts females, since males have a larger dewlap.

GREVY'S ZEBRA

The third and largest species of zebra is called Grevy's zebra. They stand about 4.1 to 5.25 feet (1.2 to 1.60 m) tall and

The stripes on a Grevy's zebra do not extend onto its belly.

weigh up to 1,000 pounds (454 kg). Grevy's zebras are the most endangered of the three species, mainly because of habitat loss due to human activity. A shrinking habitat means zebras are competing with each other and with other animals for food, water, and other resources. This kind of competition often ends in death in the wild, either because the animals fight each other to the death or because the weakest lose out on the resources they need to live.

Mountain zebras are good at climbing!

CHAPTER 2

HABITS AND HABITAT

All three zebra species can be found across the African continent. They are herbivores, which means they eat only plants. Their diet is mostly made up of grasses, sometimes **supplemented** with the stems and leaves of certain bushes. Much of a zebra's day is spent eating. The animals use their sharper front teeth to clip the ends off the grass or bush, then they grind it down with their flat back teeth.

Each of the different species lives in a different habitat, although each habitat includes grasses for the zebras to feed on. Grevy's zebras live mainly in the grasslands of Kenya, Ethiopia, and Somalia. Plains zebras, being the largest population, are widespread across multiple African countries, mostly on the eastern side of the continent. Mountain zebras live in and around the mountains of Namibia and Angola.

THAT'S WILD!

A zebra's back teeth keep growing its whole life. This is because grinding its food wears down the teeth. If they didn't keep growing, an adult zebra would not be able to eat.

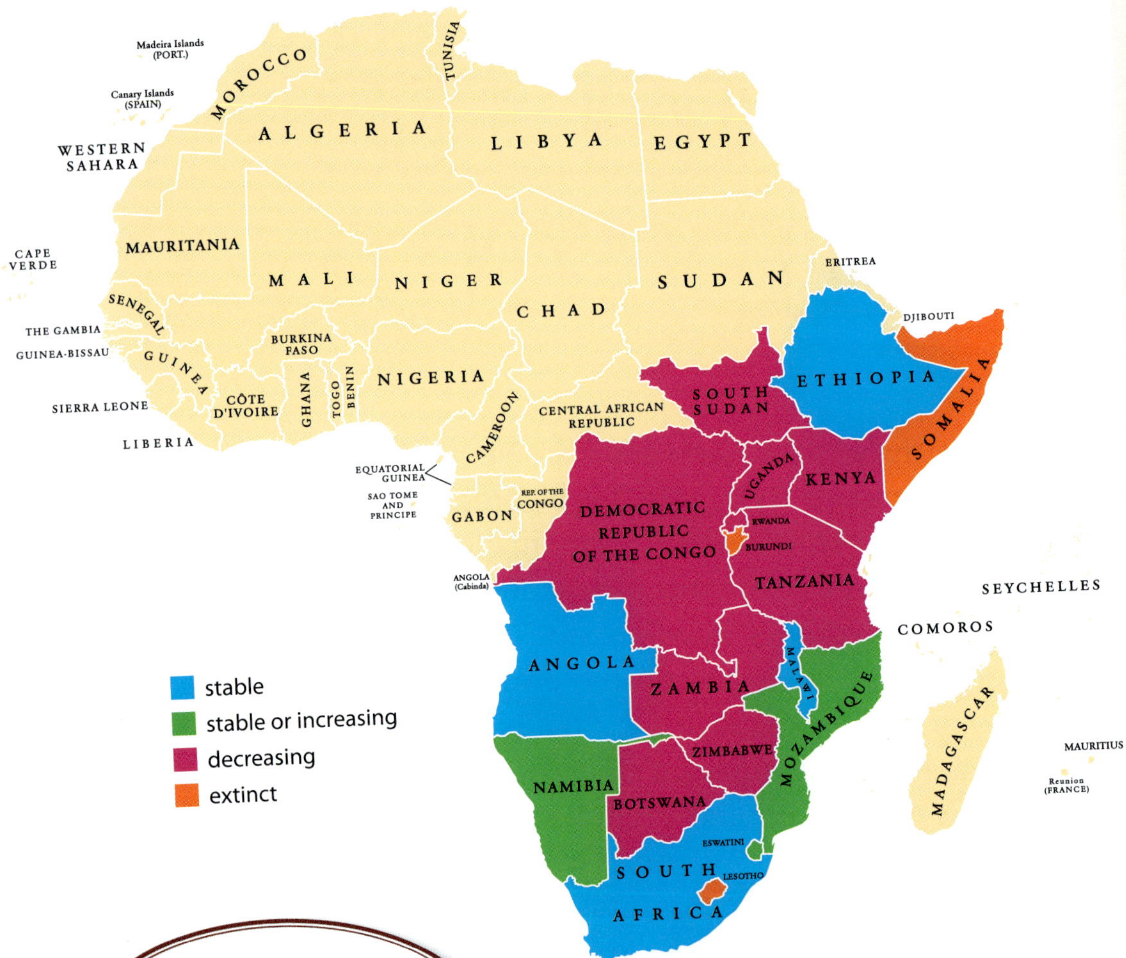

This map shows the state of the plains zebra population in different African countries. Experts believe the zebra is extinct in Somalia, but more evidence is needed to say definitively if this is true.

ZEBRA FAMILIES

Plains and mountain zebras are social animals that live in small family herds that are called harems. Each is led by one stallion,

THAT'S WILD!

Like all animals, zebras need to sleep. One zebra stands guard while others rest. A zebra can doze while standing because its legs lock in place to hold it up, but they do sometimes lie flat on their sides to sleep.

or male zebra, and generally includes several mares, or females, and their foals. Sometimes many different harems will join together to create a larger herd, but the families tend to stay together even within the herd. Plains zebras have been known to create large herds of between 80 and 1,000 members, while mountain zebras typically have no more than 15 in their group at any one time.

The stallion of a zebra family chases other stallions away from its mares and foals. A stallion also protects its family from predators by placing its body between the predator and the other zebras. One mare helps the stallion manage the family. It is often the oldest mare. This mare always walks in front of the other mares to lead them on trails, with the stallion last in line to keep an eye out for predators that might be sneaking up on the family. The mares in a family form close bonds with one another. They stay together as a family even if the stallion dies and a new stallion takes over.

Zebras have certain behaviors they only use with their families. One is mutual grooming, where two zebras will clean each other's coats at the same time by softly nibbling at the neck, sides, and back. Groups do not leave sick or hurt zebras behind; instead, they wait for them and protect them from predators. Outside of family groups, stallions sometimes fight by kicking and biting each other. This can be in competition for a mate or food, or because they are annoyed with each other. Their manes help protect them from neck bites.

NOMADIC ZEBRAS

Zebras are nomadic, which means they often travel instead of staying in a specific territory. Many places in Africa have a rainy

Zebras are often aggressive toward each other.

season and a dry season. During the dry season, water sources often disappear and plants die. The main reason zebras migrate, or move around, is to find food and water when the seasons change. One of the most famous and breathtaking migration events happens on the Serengeti plains in the country of Tanzania. During the dry season there, grasses die and watering holes dry up. As a result, up to 10,000 Grant's zebras will join together and travel at least 700 miles (1126 km) in search of food and water. It is known to be one of the largest and longest migrations of any mammal in the world.

Zebras often form large herds with other kinds of animals. Wildebeests are common companions. Herding together helps all the animals—it means more eyes, ears, and noses on the alert! Wildebeests and zebras don't compete for food; zebras eat the top of the grass, and wildebeests eat the bottom.

> **THAT'S WILD!**
> Mountain zebras also migrate, but they move shorter distances. During the dry season, they move to lower elevations to find food.

The migration is important for several reasons. First, it helps prevent the zebras from dying from a lack of food and water. Second, it decreases competition for resources in certain areas: When the zebras leave, other animals have more access to the remaining food and water. Third, the migrating zebras themselves become an important food source for predators such as hyenas and lions. Such a large herd gives the predators a better chance of catching something.

A DIFFERENT LIFESTYLE

Unlike the plains and mountain zebras, the Grevy's zebra is not nomadic and does not form permanent herds. The stallions find a patch of land and mark out borders using their dung and urine. The mares wander through the stallions' territories. When the stallion in one territory mates with a mare, the mare and the foals are allowed to stay in that stallion's territory and receive protection.

When the foals are old enough to travel, the mare will often leave the territory of its mate and live a nomadic life with its children. Sometimes they form "nursery herds" with other zebra mothers and foals. A "teen" zebra is called a filly if it is a girl and a colt if it is a boy. Fillies and colts are both allowed to walk through a stallion's territory. However, if another stallion comes near, the stallion that has marked the territory will defend it aggressively. Stallions typically only leave their territory if they have trouble finding food; then they will temporarily join a larger herd and migrate.

> **THAT'S WILD!**
> A colt becomes a stallion when it is around three and a half years old.

ZEBRA COMMUNICATION

Zebras communicate with each other through six specific noises. The meaning of the noise depends on the situation in which it is used. For example, a snort warns others in the herd that a predator could possibly be hiding nearby, such as in tall grass. In some cases, however, snorts also show happiness. A third use of the snort is to let a particular zebra know it is unwelcome in the area.

A neigh lets the herd know the danger is real and visible. Zebras make a barking sound to say hello to each other or get each other's attention. They bray like a donkey to show annoyance or anger with each other, but they also use the same sound to get the attention of possible mates or to communicate over long distances. Zebras squeal when they are hurt, and babies who are in distress wail loudly.

Zebras also use body language to communicate. This zebra is showing anger and aggression by putting its head down and pulling its lips back from its teeth.

Mares take care of their foals after they are born. A mare will not move on until the foal can keep up with it.

CHAPTER 3

LIFE AND DEATH

Because zebras are generally nomadic and live in herds that stick together, zebra foals grow up fast. Their mothers cannot leave them or the herd, so the foal needs to be able to keep up! Unlike many mammals, zebras are born with the ability to see and hear. They can walk within half an hour of being born, and they can run after about an hour. However, they are still **vulnerable**, so the herd protects them. When predators are nearby, the harem will form a tight group around the foal and its mother so the predator will have trouble getting to them, and the stallion of the harem will actively attack to protect its family.

THAT'S WILD!

A zebra's gestation period, or the amount of time it spends inside its mother's body, is between 11 and 13 months.

GROWING UP

Even though foals are fairly independent soon after they're born, they still need their mothers. The foal suckles, or drinks its mother's milk, for about three months after it is born. The mother also teaches the zebra what it needs to know. When it is done suckling, the zebra is no

Zebra foals are brown and white. Their coat darkens as they grow. Foals of some subspecies have a mane that stretches all the way down their back. They lose most of this as they grow.

THAT'S WILD!

A filly becomes a mare between the ages of 3 and 6. A zebra mare can give birth once every 3 years until it is 24 years old.

longer constantly at its mother's side. It joins the larger herd and plays with other foals.

When they are about 22 months old, the foals leave the herd completely to form a community with other young zebras. In contrast to harems, which are tight-knit, this young zebra herd has loose ties. The young zebras will move together but will often join larger herds when they stop

This picture shows a foal getting milk from its mother. A mother zebra is known as a dam. A father is known as a sire.

for food and water. When the zebras get old enough to mate, they form their own harem.

In the wild, zebras can live about 25 years. Zebras respect their elderly herd members. The migration is often led by the oldest female. Since it has made the journey many times before, it is able to make sure the herd follows a safe route that is always close to fresh water.

A LIFE OF DANGER

Zebras are a favorite meal of lions, leopards, hyenas, cheetahs, and wild dogs. The herd watches carefully for these predators. The location of the zebra's eyes help it spot danger. The eyes sit on the sides of its head, which allows the zebra to see behind it and in front of it. The eyes are also placed high

A SYMBIOTIC RELATIONSHIP

Oxpecker birds often ride on zebras' backs. Sometimes a zebra may carry more than a dozen of them—with birds hanging off its sides and belly! Zebras and oxpeckers have a symbiotic relationship, or one in which two animals work closely together and help each other. The birds clean bugs off the zebra's skin and get a meal at the same time. Oxpeckers are also often the first to sound the alarm about a nearby predator, which helps alert the herd. A zebra knows a predator is near when the birds hiss and fly away.

Oxpeckers can be very helpful to zebras, but they also sometimes cause problems. The birds mainly eat ticks, which suck the blood from mammals. Oxpeckers have been seen picking the scabs off the wounds the ticks leave behind. If they do this too many times, the wound gets infected instead of healing.

Zebras can be a mode of transportation for birds such as these.

up, so the zebra can see even when its nose is deep in a patch of grass.

Lions are a serious threat to zebras.

When zebras spot a predator, they will either run or fight, depending on the situation. They are more likely to fight if foals are being attacked. Sometimes they can seriously hurt the predator and chase it away from the herd, or outrun it completely. Other times, the predator successfully takes down one or two of the zebras.

THAT'S WILD!
A well-placed kick from a zebra is enough to kill a lion.

Some animals that eat zebras are also willing to scavenge. This means they will eat a zebra or other animal that has already been killed by a different predator. These occasional scavengers include hyenas, leopards, and even hippos! For this reason, zebras quickly move on when one of their herd members is killed. Hanging around the body invites further attacks from animals that show up to scavenge.

THAT'S WILD!

Hippos are one of the most aggressive animals in the world. Even though they mostly eat plants, they have been known to attack, kill, and eat zebras as well as scavenge them.

ALWAYS WILD

Since zebras and horses are so similar, many people want to know if it is possible to tame a zebra. In general, the answer is no. Nearly all horses alive today are domesticated. This means that over thousands of years, through selective breeding, they have developed traits that make them fairly easy to train. Zebras have not had this selective breeding. Their aggression makes them dangerous to even approach, let alone ride, and their skittishness makes them unreliable even if they have been trained.

The fact that zebras are not domesticated has not stopped people from trying to tame them. As far back as the 1700s, people have been looking for ways to domesticate zebras. European colonizers in Africa wanted animals that could carry packs and pull carts. Some succeeded in taming individual zebras, but this was done with great difficulty, and people decided over time that it was not worth the effort.

Even when they gave up on the idea of taming zebras, some people still wanted to own them. One example is William Randolph Hearst, who grew rich by establishing a

Some people were able to train zebras to pull carriages through the streets of London in the late 1800s.

newspaper that grew into a large media company in the United States. Hearst had a private zoo that contained zebras.

After Hearst died in 1951, people stopped taking care of his estate, which included his home and the surrounding lands, and the animals were let loose in the Californian countryside. To this day, there is a herd of about 150 wild zebras in California.

National parks, such as Kruger National Park in South Africa (*shown here*), have rules against hunting the animals that live within the park's borders. This is one way people are trying to stop animals such as zebras from going extinct.

CHAPTER 4

SAVE THE ZEBRAS

Today, the plains zebra is not in danger of extinction. They are listed as "near threatened," which means they are doing well right now, but their population is slowly dropping. If this continues, they could become endangered. The Grevy's and mountain zebras are already endangered. Mountain zebras are listed as "vulnerable," and Grevy's zebras are listed as "endangered." Scientists determine these categories by looking at things such as how big an animal population is; whether the population is stable, growing, or shrinking, and how quickly it is growing or shrinking; and how threatened they and their habitat are.

HUNTING

In the past, hunting was a big reason for the zebra population's decline. Today, zebras are protected by law from hunters in most places, but illegal hunters called poachers are still a problem. Poaching is generally the result of poverty. People either kill animals to eat or to make money by selling the body parts, such as zebra skin or elephant tusks. In 2021, poaching increased in Kenya because a widespread drought

THE RED LIST

The International Union for Conservation of Nature (IUCN) was created in 1964 and has since become responsible for keeping track of which species are threatened. Its "Red List" has eight categories:

- **least concern:** not at risk and not likely to be at risk in the near future
- **near threatened:** close to being at risk
- **vulnerable:** likely to become extinct in the near future without human conservation efforts
- **endangered:** high risk of becoming extinct in the near future
- **critically endangered:** nearly extinct
- **extinct in the wild:** living members of the species can only be found in zoos or specially set up conservation areas
- **extinct:** no living members of the species remain anywhere in the world
- **data deficient:** not enough information to determine a threat level

With millions of species of plants and animals in the world, it is impossible for the IUCN to have them all on its list, but more are added every year.

increased poverty and made it difficult for people to find or afford food.

Some people also hunt zebras as part of a practice called trophy hunting. This is when hunters want the animal's head or skin so they can brag that they killed a large animal. Other animals that are commonly targeted for trophy hunting

Zebras are sometimes hunted for their skins.

include lions, giraffes, black rhinos, elephants, and leopards. Trophy hunting is legal in some parts of Africa with the proper permits, and some countries allow hunters to bring their trophies home. Conservationists are pushing to have these practices outlawed, since many of the animals that are being hunted are endangered.

THAT'S WILD!

Many of the largest animals in the world live in Africa, which is why trophy hunting is more common there. In North America, brown bears and gray wolves are sometimes targeted by trophy hunters.

HABITAT LOSS

Another threat to zebras is the loss of their natural habitat. This is especially true for Grevy's zebra, which has suffered more habitat loss than any other African animal. This happens mainly because of human

This picture shows Grevy's zebras on a nature reserve in Kenya. Nature reserves are a big part of animal conservation.

encroachment. One of the biggest problems is that farmers let their cattle herds graze on land that is populated by zebras.

Farmers also drain water from rivers for their crops. Furthermore, climate change causes droughts that last a long time, further reducing the amount of water available. Zebras must compete with cattle and crops for food and water, forcing the zebras to migrate farther and farther in search of enough resources to keep them alive. Long journeys are especially hard on pregnant zebras and foals, which often die along the way.

THAT'S WILD!

Zebras are also threatened by a disease called African horse sickness, which is spread by biting insects. Most forms of the disease can be deadly.

Cattle farming is a big part of why zebras are endangered.

ZEBRAS OF THE PAST

Zebras, horses, and donkeys are all related, but distantly. They all came from a common ancestor called *Eohippus*, which lived about 52 million years ago. It was small and quick, like a fox. It lived mainly in jungles, where it could hide among the plants from its predators. Over time, *Eohippus* left the jungle and moved onto the plains. With nowhere to hide, it needed to get fast enough to outrun predators.

Over the next few million years, *Eohippus* evolved, or changed, into many different kinds of **proto-horses** that were better suited to their new environment. They grew bigger, and their five toes merged into one hoof, which helped them run faster. Eventually, they became the animals we know today.

Fossils show that zebras, particularly Grevy's zebras, once roamed from Africa all the way to China. Human threats have shrunk their habitat.

This drawing shows what scientists believe *Eohippus* looked like.

WHY SAVE THE ZEBRAS?

Saving animals from extinction is an important job. Once an animal goes extinct, it can never come back. Scientists have tried to "de-extinct" certain animals, such as the woolly mammoth, but any animal they create is not the same genetically as the animal that has gone extinct. Furthermore, creating new animals causes environmental problems, as the new species will now compete with existing species (many of which are already endangered) for important resources.

Saving zebras is valuable because zebras have important roles in their environment. Scientists call zebras "ecosystem engineers." This means they change their environment in ways that help both themselves and other species. For example, in very dry weather, mountain and Grevy's zebras will often use their sharp hooves to get at underground water. They make pits that fill with water and drink from them. When they move on, other animals use the watering hole, including plains zebras, giraffes, hyenas, and lions.

Another example of ecosystem engineering is the mountain zebra's practice of dust bathing. Zebras will roll around in the dirt, creating pits that get bigger as more zebras use them. Scientists have studied the dust in these pits and found that it holds moisture better than the surrounding soil when it rains. This means more plants grow in the dust pits, and they are often the plants zebras rely on for food.

A third way zebras help the environment is by being what scientists call "grazing pioneers." When they enter a new grazing area, they eat the tips of the tall grasses, preparing the area for other animals that only eat shorter grasses. Without zebras, those animals would die out as well. Furthermore, some of the plants zebras eat grow quickly and spread easily. Their grazing stops these plants from taking over and

Zebras take dust baths to help keep themselves cool in the hot sun and to remove bugs from their coat.

leaving no space for other plants in the ecosystem.

CONSERVATION EFFORTS

Many different groups are working on many different ways to save zebras. Some groups are working with African farmers to help them use more **sustainable** agricultural practices. Others are creating areas of land called wildlife reserves where zebras can live safely. Poachers are still a threat in these areas, but guards are posted to try to catch poachers before they can get to the zebras.

A group called Grevy's Zebra Trust has been working with women of the Kenyan Samburu tribe to monitor Grevy's zebras. The organization trains Samburu women to be zebra scouts, tracking the animals and noting things such as their health, migration routes, and threats or positive developments. The scouts also help the zebras by doing things such as digging water wells during the dry season, and they educate the rest of their tribe about the zebras' habits and the importance of helping them. This gives the scouts, who are often widows and single mothers, an important job to do as "citizen scientists." They earn money as well as the respect of other members of their tribe for their knowledge; the zebras get help; and the jobs stay with locals in the area instead of being taken by scientists from other countries.

THAT'S WILD!

In 1937, there were fewer than 50 Cape mountain zebras still alive. Today, thanks to the creation of Mountain Zebra National Park, there are several hundred. Most live within the park's borders, but a few roam outside it.

This last point is especially important because the survival of the Grevy's zebra depends on how well it can live alongside the humans who also call the zebra's natural environment home. When locals are involved in conservation efforts, the results are much better.

THAT'S WILD!

Some zoos and parks claim to help zebras, but actually harm them. Places that let visitors touch, hold, or play with zebras are not following safe practices. Experts say these actions make zebras feel stressed and upset.

THINK IT THROUGH

1. Why do you think horses have longer tails and manes than zebras?
2. Why do people want to bring the quagga back from extinction?
3. How do zebras benefit from forming both small harems and large herds?
4. What other ways might the environment be harmed if zebras go extinct?
5. Do you think trophy hunting should be legal? Why or why not?

GLOSSARY

aggressive: Showing a readiness to attack.

ecosystem: A natural community of life that includes living and nonliving features.

encroachment: The process of advancing beyond the usual or desirable limits.

pigment: A substance that gives a plant or animal color.

proto-horse: An ancestor of modern-day horses, donkeys, and zebras that had some but not all of the characteristics associated with those animals.

selective breeding: A process by which humans choose which animals will breed with each other so certain characteristics become more common.

subtle: Difficult to understand or distinguish.

supplemented: Added in order to make something complete.

sustainable: Able to last for a long time.

taper: To gradually disappear or come to a point.

unique: Special or one of a kind.

vulnerable: At risk for attack or damage.

FIND OUT MORE

Books

Hubbard, Ben. *Lions vs. Zebras: Food Chain Fights*. Minneapolis, MN: Lerner Publications, 2025.

Gish, Melissa. *Zebras*. Mankato, MN: Creative Education, 2022.

Maloney, Brenna. *Horse or Zebra*. New York, NY: Children's Press, 2024.

Parkin, Michelle. *A Dazzle of Zebras*. Parker, CO: The Child's World, 2024.

Websites

Kiddle: Zebra Facts for Kids
kids.kiddle.co/Zebra
Read more interesting facts about zebras, and take a look at some amazing photos.

Louisville Zoo: Kids for Conservation: Zebras
louisvillezoo.org/kids-for-conservation-zebras
Read about how the Louisville Zoo in Kentucky takes care of its zebras.

***National Geographic Kids*: Zebra Facts!**
www.natgeokids.com/nz/discover/animals/general-animals/zebra-facts
Learn 10 cool facts about zebras.

World Animal Protection US: 5 Fascinating Facts about Zebras
www.worldanimalprotection.us/latest/blogs/5-fascinating-facts-about-zebras
Read more interesting information about zebras.

YouTube: What Sound Does a Zebra Make?
www.youtube.com/watch?v=xPmlh_ncsko
Listen to a zebra's bark.

Publisher's note to educators and parents: Our editors have carefully reviewed these websites to ensure that they are suitable for students. Many websites change frequently, however, and we cannot guarantee that a site's future contents will continue to meet our high standards of quality and educational value. Be advised that students should be closely supervised whenever they access the Internet.

INDEX

C
California, 31
cheetahs, 27
climate change, 37
communication, 23
conservation, 34, 36, 40, 41

D
disease, 37
dogs, 27
donkeys, 5, 23, 38

E
ecosystem engineers, 39
eyes, 27, 28

F
farmers, 37, 40
foals/babies, 7, 19, 22, 24, 25, 26, 27, 29, 37

G
grazing, 37, 39
Grevy's zebras, 7, 14, 15, 17, 22, 33, 36, 38, 39, 41
grooming, 19

H
habitat loss, 15, 36, 37
herds, 4, 18, 19, 21, 22, 23, 25, 26, 27, 28, 29, 30, 31
hippos, 30
horses, 4, 5, 11, 30, 38
hunting, 9, 32, 33, 34, 35, 36
hyenas, 22, 27, 30, 39

L
lions, 4, 22, 27, 29, 36, 39

M
manes, 8, 19, 26
mares, 19, 22, 24, 26
migration, 21, 22, 27, 37, 41
mountain zebras, 7, 11, 12, 14, 16, 17, 18, 19, 22, 33, 39, 41
myth, 10

O
oxpecker birds, 28

P
plains zebras, 7, 8, 9, 11, 17, 18, 19, 33, 39

Q

quagga, 8, 9

R

Red List (IUCN), 34

S

stallions, 18, 19, 22, 25
stripes, 4, 7, 9, 10, 11, 13, 14

T

teeth, 17, 23

W

wildebeests, 21

Z

zoos, 31, 34, 41

ABOUT THE AUTHOR

Jennifer Lombardo earned her BA in English from the University at Buffalo and still resides in Buffalo, New York, with her cat, Chip. She has helped write a number of books for young adults, on topics ranging from world history to body image. In her spare time, she enjoys cross-stitching, hiking, and volunteering with Habitat for Humanity.